*A WHALING VOYAGE IN THE PACIFIC OCEAN
AND ITS INCIDENTS*

A WHALING VOYAGE IN THE PACIFIC OCEAN AND ITS INCIDENTS

George A. Dodge

Edited by Kenneth R. Martin

YE GALLEON PRESS
FAIRFIELD, WASHINGTON

1998

Library of Congress Cataloging in Publication Data

Dodge, George A.
    A whaling voyage in the Pacific Ocean and its incidents.

    Originally published: Salem, Merrill & Mackintire, 1882.
    1. Whaling—Pacific Ocean. I. Martin, Kenneth Robert, 1921-
II. Title.
SH382.6.D62   1981   639'.28'09164   81-4753
ISBN 0-87770-243-8   AACR2

Introduction

Fifty years after George A. Dodge had sailed on board the Nantucket whaleship Baltic, *the experience was still fresh in his mind, and in 1882, the* Salem Gazette *published his brief memoir. Unlike whaling accounts based upon day-by-day journals or logbooks, Dodge's is selective in content, relating only the highlights of an eventful voyage. It is thus very readable. And as a whaling memoir it differs significantly from others of its type and period, which are usually lengthy reminiscences of successful seamen grown garrulous and worldly in old age.[1] Dodge apparently read few or none of his contemporaries' accounts. He offers no morals to his story, and avoids stating even the conventional wish that readers will find his work interesting. But interesting it is. Students of whaling will find it enjoyable though unsurprising, and casual readers will find it a quick and easy guide to the heyday of Yankee whaling.*

Not much is known about Dodge. Records at the Essex Institute, Salem, Massachusetts, describe a George A. Dodge of Salem, probably the author, who was born in 1814.[2] Following his youthful experience on board the Baltic, *Dodge married Charlotte H. Coffin in 1837, though the marriage ended with Charlotte's death in 1840.[3] In 1849, Dodge took the ship* Capitol *from Boston to San Francisco, where he became associated with Naumkeag Ming and Trading Company.[4] Ten years later he returned to Salem to pursue the currier's trade.[5] Known as an "ardent Democrat" and "a man of considerable ability, and of pronounced opinions," Dodge resided in Salem until his death in 1889.[6]*

Dodge's skipper on board the Baltic, *William Chadwick of Nantucket, ended a long whaling career after the voyage Dodge describes.[7] The* Baltic, *a 410-ton ex-merchantman, had made one whaling voyage before 1831, captained by Chadwick.[8] Following Dodge's, she made two more whaling voyages from Nantucket and one from Fairhaven. She was wrecked in the Aleutians in 1846.[9]*

Kenneth R. Martin

5

NARRATIVE.

SHIPPING ON A WHALER.

N EIGHTEEN HUNDRED AND THIRTY-ONE I SHIPPED on board of the ship Baltic, of Nantucket, commanded by Captain Chadwick, a man of sixty years. Our crew consisted of thirty men, four mates and four harpooners. The custom in those days was to have an agent in Boston to ship green hands from the country, because they could be easily duped. These agents told us fine stories about whaling, how much money we could make, and what a pleasant voyage it was, and held out every inducement for us to ship. We were finally persuaded to go to Nantucket and ship. Our pay was to be one barrel of oil out of one hundred and eighty; the captain had one out of fourteen, the mate one out of fifty, and the other officers in proportion to their rank. The owners provided us with an outfit, and charged us twenty-five per cent insurance and twenty-five per cent interest on the money, so you can see that a large slice of our voyage was gone before being earned. The ship had to be taken to Edgartown to take in her supplies for the voyage.[10] After about three weeks, all things being ready, we set sail for what proved to be a long and tedious voyage.[11]

THE SHIP SAILS.

After getting out into blue water, the ship cleared up, and everything put to rights, we set to work getting the whaling gear in order for the first spouter as soon as he should show his head above water. Nothing unusual took place for a number of days. One bright morning we raised a school of black fish.[12] Now here was a chance for some fun, as we wanted some oil for the ship's use. We lowered our boats, and captured enough to make ten barrels of oil. Blackfish are a species of whales; they are not very large; they make all the way from three to eight barrels of oil. But it is sport to get fastened to one of them; they will take you through the water at the rate of two forty, and no discount. I was wrought up to the highest pitch of excitement, this being my first introduction into the mysteries of whaling. To see these creatures dart first to the leeward, then to the windward, was truly wonderful; sometimes heaving themselves clear out of the water, coming down with a splash that would send the spray all over you. After we got done trying out, our time was taken up in making spun yarn and sinnet,[13] grinding harpoons and lances, so as to be ready for the first spouter as soon as he showed his head above water.

THE FIRST WHALE.

On the twenty-eighth day out, the man at the masthead cried out, "There she blows." "Whereaway?" says the Captain. "Two points on the lee bow," answered the man from aloft. The man at the helm was told to keep the ship off two points. By this time the whale had gone down, and all hands were looking for the next rising.[14] The whale generally stays under water one hour, by the watch. I never knew it to fail, and it is said they never breathe under water. When they come up they spout once a minute, spouting sixty times, making one hour, and then go down again. When the whale came to the surface, the excitement was beyond description, it being the first whale that we ever saw, most of us being green. The whale being near enough, the captain gave orders to take in the light sails, haul up the courses, haul aback the main yard, and lower down the boats, all under one breath. No man-of-war ever presented a wilder scene of commotion when going into action than we did on board of our ship when we were about attacking our first whale. The scene when lowering the boats beggars description. One fellow sung out, "Pick me up, Tucker!" Another lost his hat in his hurry to get on to the boat. I will assure the reader that it is no easy job for a greenhorn to clamber down the sides of a ship in a heavy sea.

The boats being all down, the race began, and it was an exciting one, each officer being eager to get alongside of the whale first. "Pull, boys, pull!" was the watchword; "we will have him the next time he spouts." The fourth mate was the lucky man, having got the first iron into the whale, much to the mortification of the chief mate, he being very anxious to get the first whale. The number of whales each officer gets is recorded on the ship's book, and the one getting the highest number is made a hero of when the ship gets home, and was considered a greater man than the President of these United States. To get a wife in Nantucket, you had to double Cape Horn three times, fasten to a whale, and do everything else appertaining to a whaling voyage. If you were a good whaleman, you could take your choice of the Nantucket girls.

I shall never forget the appearance of this huge monster of the deep, as we pulled alongside of him; he looked like a large ocean steamer, paddling along with his fins and flukes. But the excitement reached its climax when the whale was made to spout blood. It takes about an hour for one to die. They live till all the blood is spouted up; when they are about to give up the ghost, they take a circle around, and heading toward the sun, they expire. If the ship is to the windward, she runs down and takes the whale alongside. A large whale will reach from the ship's fore-chains to her mizzen, which is about seventy-five feet. The whale being thus secured, we begin preparations to cut in the blubber. This is done by means of a large tackle and fall, which is made fast around the head of the mast. A staging is built over the side for the officers to stand on while cutting in. A man has to go on top of the whale to hook on to the first piece. This job belongs to the harpoon's man, as it is quite a difficult task to perform, and sometimes a very dangerous one, when there is a heavy sea. Sometimes the sharks are very annoying, and I have seen the mate slice one in two just in the act of taking a man's arm into his mouth. After the blubber is all taken from the body of the whale, the head is separated and hoisted up at the side of the ship. About twenty-five barrels of oil are taken out of a whale's head. This is done by setting a bucket into the head and hoisting it out with a tackle. This is what is called the head matter, and needs no trying. The blubber that comes from the body of the whale is sliced up and tried out. There are three large try-pots that sit abaft the windlass, and the scraps are used for fuel in trying out. After the oil is tried, it is put into casks and stowed in the hold of the ship.

ON THE PACIFIC.

After we got through trying out this whale, we set sail for Cape Horn, reaching it in about two months. We experienced very heavy weather whilst doubling the Cape, but soon got into milder weather in the calm Pacific. We ran down to the Island of Juan Fernandez,[15] sent a boat on shore, and got a good supply of fruit, which was a godsend, after living on salt grub five months. This island was inhabited by Spanish convicts, sent there for crimes. It was a lovely spot of God's earth, abounding with all kinds of tropical fruit and everything to make a man happy; and yet men were sent to this lovely spot for crime. I should consider it a blessing rather than a punishment, even if I was alone, as Robinson Crusoe was when he was cast away on this same island.[16]

The next port we made was Coquimbo, where we went to recruit for a whaling cruise. This port is in Chili. It is inhabited by a mixed race, which is called Creole. They are a very kind hearted people, and seemed desirous of entertaining Americans. Almost every person you met with would ask you into the house, and after you were seated a glass of native wine would be handed to you for refreshment. I never saw a native drunk, for they never drink anything but native wine. They were very much disgusted when they saw English and American sailors intoxicated, as was the case every day as long as we stayed there — for they were not satisfied with native wine, but must go to the dram shop and get something stronger. Our crew was no exception to the general rule of sailors, for a number of them got drunk the first day on shore, and when they came on board they had to be ironed and put down into the hold for safe keeping. You would have thought that bedlam had broken loose, after the hatches had been put on, such a hubbub as they kicked up; two darkies among them butted through all the partitions into the forecastle, and made the night hideous with their yells. The carpenter had a day's work making repairs. So the reader can see the folly of getting drunk; it not only ruins the one that does it, but makes every one around you unhappy.

Coquimbo was the first foreign port that I ever saw, and the manners and customs of the people were novel to me. In the room of shaking hands when they met, they would clasp each other around the waist, exclaiming, "Bueno fuerte Bracos." You would often see women carrying water on the top of their heads, and doing other drudgery more fitting for men. The heavy burdens were carried on the backs of jackasses. The houses are built of dried clay, cut in squares like bricks: they are called adobes; the houses are built one story high, and have nothing but a ground floor. The cooking was done out of doors, the washing was done at the brooks in cold water. The men were rather indolent, preferring gambling rather than hard work.

ANOTHER WHALE.

After spending a few weeks here, we set sail for a cruise. We had not been many days at sea before the man at the masthead cried out, "There she blows!" All hands were full of excitement, and as eager for the chase as ever a dog was for the fox. The boats were lowered, and after a long chase we captured a whale without any accident. I thought then that it was all fun to catch whales, but I found out further along on the voyage that there was oftentimes more danger than fun. We cruised about here nearly two months, taking in about four hundred barrels of oil, which ought to have been a thousand, had it not been for bad management.

THE SANDWICH ISLANDS.

Whales having become scarce in this latitude, we made sail for the Sandwich Islands to recruit, which we reached in about two months. I shall never forget the beauty of the scenery going into Mowee. These islands are very pleasant and healthy, and the natives seemed happy and contented, and they had nothing to make them otherwise. Everything grew spontaneously there. Bread-fruit grew on trees; yams, sweet potatoes, plantains and bananas, grew without much cultivation; and in fact there was hardly any labor to be performed.[17] The natives were almost in a state of nudity, wearing only a breech cloth made of the bark of a tree. I have often seen them going to a missionary meeting on Sunday with nothing but a hat on, that some sailor had given them; another with a swallow-tailed coat and brass buttons; another with a pair of boots and spurs, looking more like a gamecock than a man or woman. There was nothing to distinguish the sex. Imagine for a moment a man going through the streets of Salem dressed in such a garb, and you can form some conception of how a Sandwich Islander looked fifty years ago when the missionaries first began their labors among them.

A roasted dog was a great dish, and eaten with as much gusto as we would eat turkey. The natives would chew up sugar-cane, spit it out and ferment it, and then drink it for liquor. There was no necessity for prohibitory laws, for as long as they had good grinders they had home supplies; but they soon learned to drink American rum after we commenced exporting it to these islands. I have seen them eat a raw fish, ommencing at his tail, and making a sumptuous dinner. I have also seen them pick the fleas off a dog's back, and devour them as quick as a monkey would. This astonished me more than anything I saw; for they had enough of everything to eat, and the most delicious fruits.

MORE WHALES—THE WRITER HURT.

We stopped here about two weeks, taking in water, sweet potatoes, and other luxuries, for a five months' cruise in the Chinese Sea. After being at sea for some time without seeing a whale, one fine morning,—the sea being as calm as a mill pond, the deep blue sky over our heads gave us the appearance of being in a little world by ourselves, there being no whales in sight of the naked eye,—the mate being anxious to catch sight of a spouter, took his spy-glass and went aloft. After scanning the horizon for a few minutes, he saw a spout, and immediately cried out, "There she blows!" Everything now was noise and bustle, awakening us to new life and activity. There being no wind at the time, it was impossible to get the ship any nearer the whales, consequently we had to lower the boats for a long pull. After pulling about ten miles, we came up to the spouters. It was necessary to make as little noise as possible, so we peaked our oars and took our paddles. Not a ripple was on the water, nor a sound to be heard, except the spouting of a large eighty-barrel whale, stretched out upon the water basking in the sun. We paddled up to him, and before he realized any danger, the harpoonman had two irons into him chock to the hilt, setting him to spouting blood at once. This so crazed him that before we could shift the paddles and take the oars to back water, the old fellow was after us with a sharp stick. He had his head under water, and his tail fanning fore and aft, feeling after the boat. He knocked all the oars out on one side, broke three streaks out of her, and the thwart that I sat on. The mate and some of the crew jumped overboard. I was the only man hurt; he knocked an oar against my back, which rendered me insensible, and when I came to they were taking me on board the ship in another boat. The other two boats fastened to the whale, and succeeded in capturing him. He had to be towed to the ship, which took until midnight. Thus ended an arduous day's work; if the whale had struck down on the boat with his flukes, he would have sent us on a visit to the heathen Chinese.

I was laid up about three weeks, and the first time I was able to go in the boat again, we got capsized. The ship was under close reef topsails at the time; we saw the whales going to leeward at the rate of ten miles an hour; we lowered the boats, set the sails, and pulling with all our might, got alongside of a monster of the deep just as he was turning his flukes. The harpoonman planked one iron into him, and as he was going down, he capsized the boat. The harpoonman put his foot on the weather gunwale, and sent the second iron into him as the boat was going over. Here we were in mid ocean, with no other boat near us. Clinging to the bottom as the boat went over, I got a lance warp around my leg, which kept me under the boat. I remained in this perilous position until a colored man by the name of Freeman came down under the boat and cleared me. We both came up together spouting like two whales. After we got on the bottom of the boat, every sea that came would knock us off. I was almost exhausted when the other boats came to our relief. In the meantime the whale had fled, with two irons in him and a hundred fathom of line. Thus ended this day's work, with a wet jacket and no fish. It was a great disappointment to us to lose this whale, for we had been quite unsuccessful in getting oil, and this was a very large whale.

THE SOCIETY ISLANDS.

We cruised a number of months in this latitude without meeting with any more disasters, taking in about three hundred barrels of oil. We then made sail for the Society Islands to recruit, and to get rid of the scurvy, which some of our men had very badly. We arrived at Uhana[18] after a short and pleasant passage. This was a beautiful island, and I shall never forget the loveliness of the scenery as we sailed up the harbor. The natives were similar to the Sandwich Islanders, and they seemed to be perfectly happy, and in fact they had nothing to make them unhappy, for they had little or no labor to perform, and their wants were few. All kinds of tropical fruit grew in abundance, and there were no bolts or bars to keep them from helping themselves. Their huts were built of bamboo, thatched over with leaves; they lived in common, and the most radical communist might be satisfied without a division of property. The women shave the hair off the top of their heads, then filling it full of cocoanut oil, stick it up all around their heads, making it look more like a porcupine's nest than anything else. I have often seen these women miles from the shore, playing in the breakers and bathing in the sunshine, like so many fish, fearing no danger, nor worrying about what they should have to eat or drink. They would often swim to the ship, climb up her sides, and dive into the water like so many boys. I think they were the happiest people that I ever saw. It was a lovely spot of God's earth to live on in their native purity, but time has wrought great changes there since. They have become civilized, and adopted the manners and customs of civilized nations.

ANOTHER CRUISE.

After taking our supplies of vegetables and water, we set sail for another cruise. The next day out of port we fell in with whales, and captured three small ones, making forty barrels apiece; these were cow whales, which always go in schools with their calves, and never make so much oil as the he whales. There is always one he whale with the school, and when they meet another he whale without a family, the two have a regular pitched battle, and the one conquering takes the family, the other having to go in search of another, if by chance he can fall in with a family of whales where he can whip the father of the flock. It was some time after before we saw any more whales.

I will here relate a little incident that happened. A flying-fish flew from the water into the frying pan, as the cook was getting his dinner. This may seem incredible to the reader, but, as the old Methodist preacher said, paradoxical as this may appear unto you, O, Lord, it is nevertheless true.

Another incident of the voyage was this: The captain ordered the steward to have some salt beef that was spoiled cooked into mince pies for the cabin. The brine had all leaked out of the barrel, consequently it was unfit for use; but the captain, being a very economical man, was determined that the meat should not be wasted. But in this he was foiled; for all hands swore by all that was good and great they would not eat stinking meat pie, or no pie, so overboard it had to go, and a madder man you never saw than the captain. He said that there was no difference after you had eaten a thing whether it was sweet or sour; but we didn't agree with him, but considered it a matter of taste altogether.

Perhaps I may as well give the reader an account of the food that whales live on. A sperm whale lives on a fish called squid; they are very large, and are like jelly. A right whale lives on brit; it is found in large quantities floating on the top of the water; the whale scoops it up with his tail, and then sucks it in.

MISSING A PRIZE.

It is customary on these cruises, when you see another ship, to speak her, and if she is just from home, go on board, the captains being on one boat and the mates on the other. This gives us a fine opportunity to beg needles and thread, also tobacco and pipes. Sometimes whales are raised right in the midst of having a good time; then you see fun ahead. All friendship is forgotten for the time being, each being eager to catch the coveted prize. You see some very fine pulling when two boats from different ships are after the same whale. Instances have been known where two boats have got to the whale so near together that the harpoonman in one boat has darted his iron over the bows of the other boat and secured the whale. This was done once by an American over an English boat. One day, while in company with a ship, we raised a spouter. We lowered our boats, being nearer the whale than the other ship. The second mate pulled alongside of the whale, and the harpoonman, being a timid man, threw both of his irons overboard, not touching the whale, which was so near that he overturned the boat, but without doing any damage other than causing a wet shirt. The next time that whale came up he was so near the other ship that they lowered their boats and took the prize that we lost by a chicken-hearted harpoonman. This was a great loss to us, as the whale stowed down one hundred barrels of oil, being worth three thousand dollars—a very good day's work for Sunday. Many were the curses on the poor fellow's head for missing such a good chance; for there was no excuse; the whale lay like a log on the water, and there was but a little sea at the time.

This is only one of the many incidents that happened which tended to prolong our voyage and keep us away from our friends. It was unfortunate for us that the second mate's boat got alongside of the whale, for either of the other boats would have captured him. We cruised two or three months after this, taking an occasional whale. When there was nothing to do, we spent our time in making nick-nacks out of shells and whale's teeth. We would often catch fish when we were trying out, put them in the skimmer, and cook them in the oil, which is as sweet as lard when it is new. It was not an uncommon thing to take a steak from the whale after the blubber was taken off; the meat is of a coarse grain, but sweet and good. We would often fry a mess of doughnuts in the same way.

GEORGE A. DODGE

A VISIT TO TOMBUS.

Having been out some time, and getting short of water, the captain concluded to go to Tombus[19] and recruit. This was a small inland town on the coast. Nothing of interest occurred during our passage. We arrived there after a short passage. We lay out in the stream, and we had to raft our water across the bar through the breakers. This was very dangerous, and a great many boats have been upset on this bar, and some of their men have been drowned. I had more fear crossing this bar than of all the whales that I had been alongside of. After we got our water on board, the captain took a boat's crew and went up a fresh water river to the town. I was one of those selected to go. It was no desirable task, for we had to pull a long distance on a very hot day, and the mosquitoes were so very annoying that I thought there would be nothing left of us but our buttons when we reached the place of our destination. The shore was lined with alligators and other venomous reptiles. After a long pull we arrived, with what little blood we had left, glad enough to lay our wearied bodies down to rest. Here again we were disappointed, for we had no mosquito bars, and you might as well have tried to sleep in a hornet's nest.

If ever a poor sinner was glad when morning came, it was your humble servant. The place was inhabited by a few Spanish farmers, who were well disposed people, and seemed to have enough of this world's goods to make them happy, if it were not for the unruly mosquitoes. One night was enough for us, and at early morn we filled our boat with oranges, bananas, and sweet potatoes, and started for the ship, having to go over the same route again, encountering swarms of those musical insects that had tormented us during the night, and seemed determined to keep it up through the day.

MORE WHALES—BETTER LUCK.

We arrived at the ship about night, and were glad enough to get rid of our tormentors. We had a right good sleep that night, and came out the next morning as bright as though nothing had happened. The ship was soon got under way, and we bid adieu to Tombus and our tormentors. After getting out into the deep blue sea once more, our masthead was manned, and everything got ready to tackle the first spouter that should show his head above water. We cruised a number of days without seeing any whales. One fine morning, just as the sun was peeping above the horizon, with a smooth sea, the captain, walking the quarter-deck dressed in a duck frock and trowsers, waiting patiently for a sound from aloft, — at last it broke upon our ears in thundering tones, "There she blows! right abeam!" cried the man at the masthead, "going to the leeward. Quick! keep the ship before the wind, and keep her steady!"

Everything now was noise and confusion, getting the boats ready to lower, and seeing if everything was all right for action. The ship having got near enough to the whales, the captain ordered the courses to be hauled up, the mainyard laid aback, and the boats lowered. Now comes the tug of war. "Pull for your lives, boys! If you get the whale I will give you everything but my wife. Pull, boys, we will have her the next rising!" The excitement is up to fever heat when chasing whales. The whales got the run of the boats, and every time we got near them they would go down and come up in some other place. After pulling four or five hours, completely worn out by fatigue and hunger, I peaked my oar and told the mate I could not pull another inch. He laughed at me and said I had calved, but gave orders for the rest to peak their oars; and they didn't wait for him to tell them a second time. The rest of the officers in the other boats, seeing us with our oars apeak, followed us also. There we lay on our oars, Micawber like, waiting for something to turn up. As soon as we stopped pulling, the whales lost the run of the boats, and came towards us. The mate told the harpoonman to stand up, and throw his harpoon as quick as they got within darting distance. One whale came across the bows of the boat, and all the harpoonman had to do was to plug him with two irons before he knew where he was or what hurt him. But the way he carried us through the water for a half hour was a caution to all whalemen. It is fine sport to be carried at the rate of "two-forty" behind a spouter—a sleigh-ride is no comparison. Sometimes a whale rounds out two or three hundred fathoms of line, and you have to look out that he don't come up under the boat. Instances have been known where they have done this, and knocked the boat twenty feet in the air.

After a while we hauled up alongside of the whale and lanced him, setting him to spouting blood. After this is accomplished, all you have to do is to keep at a distance, and let him die, which takes about an hour. We got three whales that day. Then it was my turn to laugh; for if I had not peaked my oar, the whales would have kept the run of the boats, and we might have chased them until doomsday without success. We were out all day without anything to eat but hard bread, having lowered our boats before breakfast. Thus ended this day's work, and glad enough were we to fall into the hands of Morpheus when it became night.

THE GALAPAGOS ISLAND—TERRAPIN.

After cruising about two months, we made sail for the Galapagos Islands for terrapin. They are found in large quantities on these islands, and are very rich for food. Some of them are so large as to take four men to carry them to the boat. They live on prickly pears, which are found in great abundance on these islands. We lost a man on the island, and couldn't find him; he probably got overheated and died from exhaustion. We searched for him a number of days, and then gave it up.

After getting a good supply of terrapin, we left for the cruising grounds once more. After getting to sea, and everything cleared up, a controversy came up between the captain and the men about the cooking of the terrapin. The captain was determined they should be cooked his way, and the sailors were just as determined they should be cooked as they wanted them. But talk was of no avail with the captain, so we had to resort to other means to gain our object. Every night in our watch on deck, one poor terrapin would lose the number of his mess by being thrown into the sea. This soon brought the captain to his bearings; he thinking it would be better to have the food cooked palatably than lose the whole by having it thrown into the sea. We had no more trouble about the cooking, and fared sumptuously as long as the terrapin lasted. These creatures will live six months without eating or drinking. They are provided by nature with a bladder to carry their water, and they drink enough in the rainy season to last them through the dry. They make a very palatable dish. Their fat is yellow, and as sweet as butter. They lay their eggs in the sand, and hatch them in the sun.

ANOTHER WHALE.

After cruising a few weeks, we fell in with a whale, lowered our boats, and fastened to him. He proved to be a hard customer; he made for the boat with all of the venom of his nature, with his mouth open, showing a row of teeth five or six inches in length, and we had to use the utmost exertion to keep out of his way. He managed to bite the steering oar in two, but lucky for us it was not the boat. We finally captured him, and he stowed down about eighty barrels of oil.

VISIT TO MONTEREY AND CALLAO.

The next place we visited was Monterey, California. At one time this used to be quite a place for shipping hides and tallow to the States. This was when it belonged to the Mexican government. The native Californians are well-built men of more than average height, and very good at hunting wild cattle. They can throw the lasso with wonderful precision, and fetch their game down every time. The arts and sciences were unknown here, except the art of branding a bullock with a hot iron. Each man had his own brand to his cattle, they being let loose, again to roam over the fields and mountains.

We took in wood and water, potatoes and beans. We took in sixty bushels of beans, the finest I ever ate; so after we got to sea we had beans and rice one day, and rice and beans the next. We had to dig our own wells for water, and cut down the trees for our wood, which we had perfect liberty to do without money and without price. Oysters were plenty, and cost nothing but the digging. The only amusement we had was hunting for shells on the beach, and dancing with the Spanish girls on the ground floor of their adobe houses. After two weeks we bid good-bye to the natives and the land which has since proved to be such a blessing to the United States.

MORE WHALES—CALLAO.

After getting out to sea again, and everything put to rights, the men were sent aloft to look for whales. Three men are aloft at a time, taking their regular turns of two hours at a time, and two hours at the helm. We generally had one watch on deck and one below when on the cruising grounds, excepting when we had whales; then all hands were employed trying out. We cruised about five months, taking some four hundred barrels of oil, and then made sail for Callao, Peru. We had a fine run down to this port, reaching it in safety. The natives here are about the same as in all South American ports, rather indolent and very fond of gambling. A large part of Callao was once destroyed by an earthquake. One day when I was on shore, I was very much amused to see the military drill. The companies were of a mixed race, white, black, and Indian. The officers beat the discipline into them with the butt end of a musket; and it isn't surprising to me that the Chilians have thrashed them so badly in their recent war. The frigate *Potomac* was lying in Callao at the time, having just returned from a cruise among the Malays. She was sent there by the United States government to chastise them for massacering a Salem brig's crew. She was commanded by Commodore Downs.[20]

A HOGGER CURED.

A novel scene took place in this port; some of our men told the Commodore's gig's crew that our mate had cats[21] and flogged the men with them. They swore if they caught him on shore they would flog him, and they were as good as their word. One day they met him in a ten-pin alley, and tied him up to a post and gave him a round dozen. As soon as he got clear from his persecutors, he lost no time in hunting up Commodore Downs and acquainting him with all the facts. The Commodore summoned the gig's crew on board the frigate for trial; he also summoned some of our men as evidence in the case, but they would not testify against the gig's crew; so the Commodore adjourned the court to meet at the place where the flogging was done. He then called for the testimony of the keeper of the saloon, but he, through fear of having his house pulled down over his head, would not testify. The Commodore, seeing he could get no evidence to convict the gig's crew, told the mate to go on board of his own ship and behave himself. Thus ended a trial that was a lasting benefit to us; there was no more flogging on board of our ship the rest of the voyage.

A LONG CRUISE.

We shipped two harpoonmen here to take the place of two poor ones that we shipped at home. They proved a valuable acquisition to us, as they were first-rate whalemen. Having got in all our supplies, we again made sail for a long cruise. After two or three days, we raised whales, and were very successful, taking about seven hundred barrels in ten weeks. Everything went on smoothly, and everybody was as happy as they could be under the circumstances. People are generally happy when they are successful, but when the dark days of adversity come they are gloomy and despondent; especially is this the case on shipboard; being three or four years on one ship, seeing the same old familiar faces, without any material change for months at a time, it is no wonder that they sometimes get disheartened and sometimes give up in despair. Such is life, and we have to take the bitter with the sweet, on the land or on the sea.

The whales having become scarce, we shaped our course for the off-shore grounds.[22] Sailing along one fine morning, we raised a school of cows and calves. They were going very fast, and we lost no time in getting our boats into the water. The third mate's boat got alongside of the whales first, and owing to some mismanagement the harpoonman missed the whale. The mate talked to him very severely, and hard words passed between them; the harpoonman finally told the mate to put him alongside of the whale again and he would strike her if she knocked him into hell the next minute. Accordingly they pulled up to the whale again, going to the windward of the whale when they should have gone to the leeward, so as to have been off easy. The harpoonman struck the whale, and that was the last of poor Lewis; for the whale stove the boat and swept them all into the deep blue sea. The harpoonman was the only one lost, and whether he was knocked into hell or heaven I have never learned, but presume his body was made food for the fishes. Fortunately for me, I was laid up with a lame hand, and was unable to go in the boat at this time.

After this accident, we were unable to man but three boats. We had a man in the second mate's boat who was afraid of a whale, and when getting near one could not pull a stroke; and as I had no boat to go in, I took his place for the remainder of the voyage. He was glad enough to seize the opportunity of staying on board the ship and letting me take his place in the boat. We had another man who was also afraid of a whale, and he used to save his allowance of grog when it was served out and drink it all at once when we raised whales, so as to give him rum courage. Such men are to be pitied rather than blamed, as we are not all constituted to fill every duty in life that we are called upon to perform.

After this accident things went on smoothly for a while, taking an occasional whale and keeping the ship in order. We had been out a long time on the cruise, and it was necessary to give all hands a glass of fire water every morning to keep the scurvy off. This is a disease that often proves fatal on going ashore after a long cruise; sometimes the smell of the land will prove fatal in extreme cases. This was our last cruise; the ship was leaking badly, and the captain concluded he would go into port and recruit for home. Accordingly we set sail for Valparaiso, on the coast of Chili. All hands were in high glee over the thoughts of getting home once more. Soon after shaping our course for port, the man at the masthead cried out, "There she blows! three points off the weather bow!" "Keep the ship close to the wind," said the captain, "and get the boats in readiness to lower as soon as he rises, for he is a large whale, and by his appearance looks as though he might be a hard customer to handle. Let every man do his duty and capture the monster, as we are homeward bound, and this whale will add quite an amount to our cargo." "There she blows! close to the ship. Lower away your boats!" and as quick as a flash the boats were lowered, and every man was stretching every nerve to capture the prize. "Pull away, boys!" says the mate; "we will have her the next time she spouts. She is an ugly customer, but we must take her at all hazards." The mate told the harpoonman to stand up, "and when I tell you to dart, give it to her solid." He obeyed the command of the officer, and sent the irons chock to the hilt, which so enraged the whale that he struck a bee-line for the boat. "Stern all, boys, for dear life! or he will knock us all into the briny deep to make food for the sharks! He is right on to us, boys, with his huge mouth open, ready to swallow us up, or send us flying in the air!" The whale finally reached the boat and grappled the steering oar; and by throwing the steering oar overboard, we got clear of the whale, and the other two boats captured him.

END OF THE PACIFIC WHALING.

This ended our whaling in the Pacific. After the whale was put in and tried out, we sailed for port. Nothing of importance took place on our passage to Valparaiso. This place is inhabited by the same class of people as the rest of the ports on the Pacific coast, and so similar it is hardly worth while to describe them again. But the place is all that a sailor could wish for what he calls enjoyment on shore. The three most important places for the amusements of sailors are the Fore, Main, and Mizzen, Tops. These were places of great resort, and sailors could get all their wishes gratified for a small sum of money.

HOMEWARD BOUND.

After staying in port about two weeks, having taken in our home supplies, the captain ordered the ship got ready for sea. The sails were loosened, and the anchor hove up with a will, which showed plainly that the boys appreciated the situation of being homeward bound, after a four years' cruise among the sperm whales of the Pacific. We were soon on our way to Cape Horn, bidding adieu to the calm Pacific, and hoping soon to be in the boisterous Atlantic; but we had to encounter once more the difficulties of doubling Cape Horn, which at best is a stormy place. Even in the summer months it snows and blows tremendously, and the rigging and shrouds are frozen with ice. You are never sure of a full watch below; you are liable at any moment to be called upon to turn out and reef topsails. Many have been the times when I have been aloft that I have had to beat my hands to keep life enough in them to work; it is one hand for the owners and the other for yourself under such trying circumstances.

After getting safely around the Cape, we got becalmed near Terra del Fuego, with a strong current setting on shore. This gave the captain a great deal of uneasiness; he paced the deck the live-long night, thinking he might at any moment go ashore among the natives, who are said to be a very dangerous race of people. When morning dawned, we were close to the land, on a lee shore; all hands stood gazing at the land, thinking of the dangers that we should meet if we went ashore, when suddenly a light breeze from the land sprung up, and we made all sail on the ship, and to our great joy we were soon out of harm's way, and making about ten miles an hour towards our own dear native land, which we all longed for and hoped to see before many months.

As every day now brought us nearer home, everybody was happy and contented. Nothing occurred to mar our happiness until one day we saw a sail in the distance. As we neared her she proved to be a long, low, black-looking schooner, having every appearance of a pirate, and steering right for us. The captain put the ship before the wind and crowded on all sail to get out of her way. Everything was got in readiness to give them a good reception in case they caught up with us. She gave us a good chase, but at night we altered our course, and in the morning she was nowhere to be seen. The ship was put on her course again, and all hands rejoiced that we were clear from the suspected pirate.

Nothing of interest occurred after this until crossing the gulf stream. A heavy squall struck us, carrying away our quarter boats, top-gallant sheets, and some of our light sails. I jumped aloft to furl the main-royal, but in the confusion they neglected to settle the yard down on the cap, and all I could do to make them understand me and let go of the halyards, was unavailing. They were calling out for me to come down, or I would go overboard with the mast. I took hold of the shrouds and slid down on to the crosstrees, where I was perfectly secure. After the confusion was over, and the yard was settled down, I went up and furled the sail. After I got on deck I was met with the exclamation, "We never expected to see you on deck again!" But I had no notion of losing the number of my mess so near home.

After the squall was over, and everything put to rights, the old ship was gliding along on her course as though nothing unusual had happened. After getting on soundings, the mate attached a hook to the sounding line, and hauled up some fine large codfish, which were a great luxury, being the only ones we had tasted for forty-four months. They found a ready market fore and aft, and were eaten with much gusto.

THE LAST SCHOOL OF WHALES.

A few days after, we raised a school of whales. The ship not being full, the captain concluded to give us one more chance to have a little fun with the black skins. All three boats got fast, and the way we were carried to the windward was a caution to old whalers. The boats were going so fast that the water was above the gunwales, and you had to turn around back to get breath. We had to put drags on the line to hold water, so as to tire him out. Drags are a square piece of board, with a becket on it, and bent on to the line. With all the drags, their speed was not impeded, and it seemed as if they were determined to run us back into the Pacific Ocean. After awhile they slackened their pace, which enabled us to haul up alongside and give them one or two lances, which made them spout blood, and made us stern all for our lives, for he was coming toward us with his jaws open, and a long row of ivory glistening in the sun, but we managed by sharp engineering to keep the boat out of his way. At last the whales headed to the sun, and gave up the ghost.

This ended our whaling, and we carried the oil into port on deck. These three whales made one hundred and twenty barrels of oil, worth thirty dollars per barrel, making in the whole thirty-six hundred dollars — a very good day's work for the last one. This sum made our voyage up to twenty-four hundred barrels.[23] After the whales were cut in and tried out, the ship was again put under all sail for port. After a few days, coming on deck from my watch below, I told the mate I was going aloft to see the land. When I reached the foreyard, Long Island was all in sight. "Land ho!" said I. "Whereaway?" said the mate. "Right ahead," said I. They laughed at me, and said I was mistaken; but on coming aloft with the spy-glass, they found I was correct.

GETTING HOME.

A happier fellow than I was, after a long and tedious voyage, would be hard to find. To see my own dear native land once more thrilled me with joy inexpressible. We expected to get into Edgartown that night, but we were destined to be disappointed. The wind died away to a calm. Along towards night a light breeze sprung up; all sail was made on the ship, and the yards squared for port; but the wind kept hauling ahead until it blew a gale, and before we could get the topsails reefed, the mainsail and foresail were blown into threads, and it was with great difficulty that we got the ship hove to under a close reefed main topsail. The wind fairly howled through the rigging, and everything betokened a serious night. This was one of the heaviest gales of the voyage, but the old ship gallantly rode out the gale, and at early morn the wind hauled in our favor, and the storm having blown its blast, we had a fine run into Edgartown,[24] where we discharged our cargo of twenty-four hundred barrels of sperm oil, fetching at that time thirty dollars per barrel, making a very good voyage for the owners and officers; but the crew, only getting one barrel out of one hundred and eighty, left but a small margin after our fitting-out and incidental expenses had been deducted. The captain had what is called a slop-chest, with all kinds of sailor's clothing for sale, and when we got out we had to pay double price for them. If we lost a hat overboard, if it was ever so calm, he would not lower a boat to pick it up, but would laugh, saying, "I have plenty to sell." Some of our crew were in debt, they had drawn so heavily from the slop-chest. They had no other way of getting money when they were in port but to draw clothes from the captain and sell them on shore for half what they paid for them.

After settling up my voyage, I had one hundred and twenty-five dollars, and I have never felt richer since, and never expect to. I bought me a trunk and wardrobe, and then took passage in a sailing vessel for Boston. Reaching the city, I had to wait until afternoon before I could get conveyance home. There were no railroads in those days, and we had to take the old Eastern stage coach over the turnpike road, arriving home about five o'clock P.M. I was greeted cordially by my friends, who expressed hopes that I would not go whaling again, but would choose some occupation on shore. I followed their advice, and have never been sorry. Thus endeth my narrative of a whaling voyage in the old ship Baltic.

NOTES

1. See for example William M. Davis, *Nimrod of the Sea; or The American Whaleman* (New York: Harper & Brothers, Publishers, 1874); and William C. Paddock, *Life on the Ocean or Thirty-five Years at Sea* etc. (Cambridge, Mass.: The Riverside Press, 1893).

2. Clipping: *Salem Gazette*, 21 January 1889, Scrap Book (Bibliographical &c.), vol. 42 (James Duncan Phillips Library, Essex Institute, Salem, Massachusetts, hereafter JDPL/EE).

3. *Vital Records of Salem, Massachusetts to the End of the Year 1849* (6 vols.: Salem, Mass.: The Essex Institute, 1916), III (*Marriages*), 302; V (*Deaths*), 211.

4. Clipping, *Salem Gazette*, 21 January, 1884, Scrap Book, vol. 42 (JDPL/EE).

5. *Ibid.*; Sampson, Davenport & Co. *The Salem Directory, 1866 containing the City Record,* etc. (Salem, Mass. George M. Whipple & A.A. Smith, n.d.) p. 62.

6. Clippings:*Salem Gazette,* 21, January 1889; *The Salem Observer,* 2 February 1889, Scrap Book vol. 42 (JDPL/EE). For a contrasting view of Dodge's identity see Frances Diane Robotti, *Whaling and Old Salem: A Chronicle of the Sea* etc. (Salem, Mass.: Newcomb & Gauss Company, 1950), p. 47.

7. Federal Writers' Project of the Works Progress Administration, *Whaling Masters.* American Guide Series (New Bedford: Old Dartmouth Historical Society, 1938), p. 72.

8. Alexander Starbuck, *History of the American Whale Fishery from its Earliest Inception to the year 1876* (1878; reprint 2 vols., New York: Argosy-Antiquarian Ltd., 1964), I, 254-265.

9. *Ibid.,* pp. 318-319, 356-357, 400, 401.

10. Large vessels found it risky to cope with an obstructive bar at the mouth of Nantucket harbor, and often resorted to Edgartown, on nearby Martha's Vineyard (Charles Boardman Hawes, *Whaling* [Garden City, N.Y.: Doubleday, Page & Company, 1924]. pp. 92-93; Clifford W. Ashley, *The Yankee Whaler* [1926; 2nd ed. reprint New York: Halcyon House, 1942], p. 33).

11. The *Baltic* departed Edgartown on 20 September 1831 (Starbuck, I, 284-285).

12. Pilot whales.

13. Spun yarn is a line made by twisting rope yarn; sennit is braided spun yarn (Captain Howard Patterson, *Patterson's Illustrated Nautical Encyclopedia* [rev. ed.: Cleveland: The Marine Review Pub. Co., 1901], pp. 216, 236).

14. A sperm whale.

15. Mas a Tierra.

16. Alexander Selkirk, the factual prototype for Defoe's Robinson Crusoe, had been a castaway on uninhabited Juan Fernandez.

17. Dodge is describing the Hawaiian royal capital, Lahaina, Maui. The town was a center of American missionary influence and a major port of call for Pacific whaleships.

18. Huahine.

19. Tumbes, Peru.

20. Dodge refers to an incident involving the *Friendship* of Salem. For a popular account of Commodore John Downes's cruise in the *Potomac*, including a punitive expedition against Sumatran Malays of Kuala Batu, see Francis Warriner, A.M., *Cruise of The United States Frigate Potomac Round the World, during the Years 1831-34* etc. (New-York: Leavitt, Lord & Co.; and Boston: Crocker & Brewster, 1835).

21. Cats-o'-nine-tails. The men of the *Potomac* were hardly strangers to flogging (*ibid.*, pp. 265-266.)

22. A ground southwest of the Galapagos Islands (Elmo P. Hohman, The American Whaleman: *A Study of Life and Labor in the Whaling Industry* [New York, London and Toronto: Longmans, Green and Co., 1928], p. 149).

23. The official tally was 2,322 barrels of sperm oil (Starbuck, I, 284-285).

24. The *Baltic* arrived on 29 April 1835 after a voyage of more than forty-two months (*ibid.*)

Index

COLOPHON

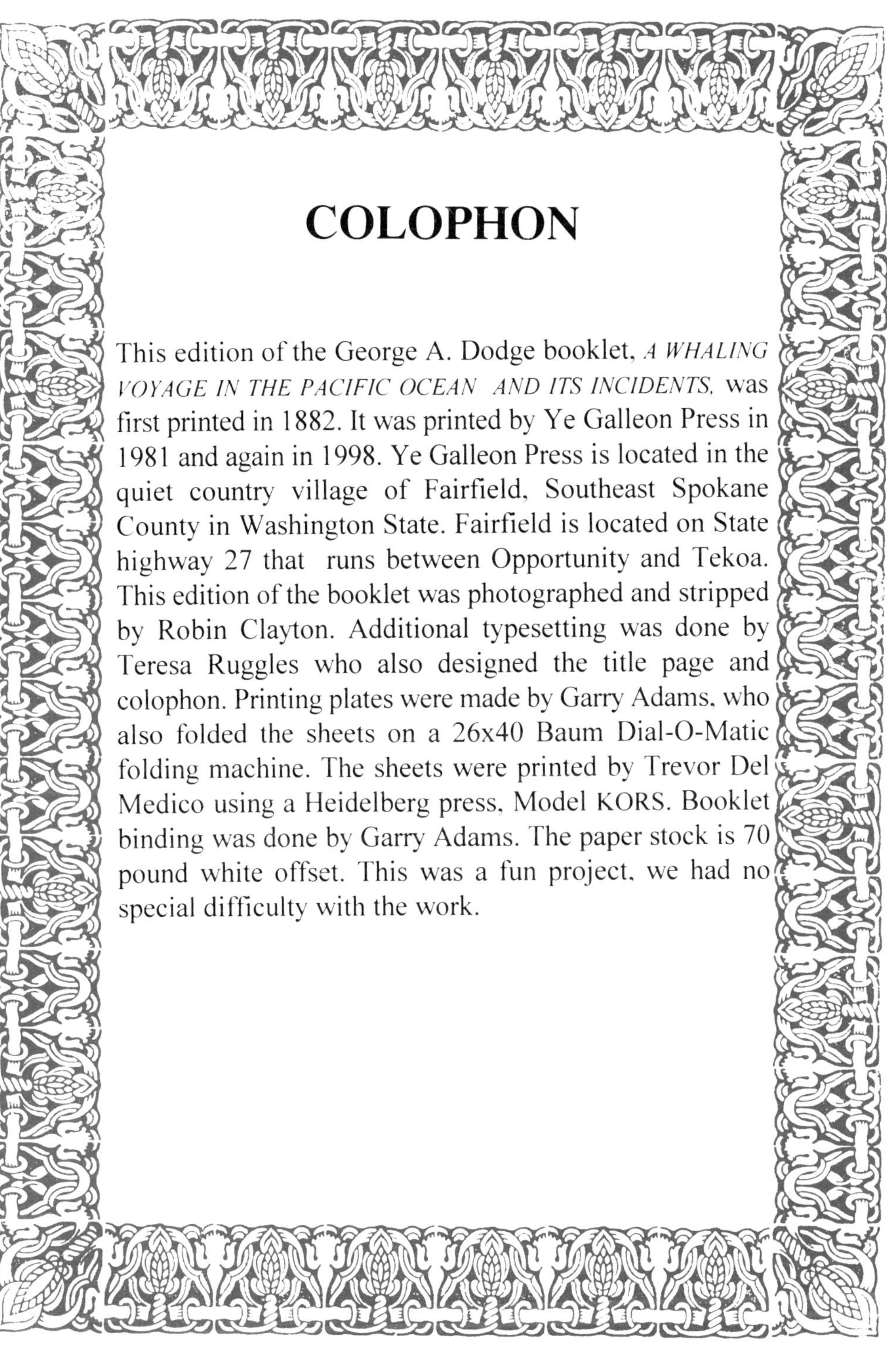

This edition of the George A. Dodge booklet, *A WHALING VOYAGE IN THE PACIFIC OCEAN AND ITS INCIDENTS*, was first printed in 1882. It was printed by Ye Galleon Press in 1981 and again in 1998. Ye Galleon Press is located in the quiet country village of Fairfield, Southeast Spokane County in Washington State. Fairfield is located on State highway 27 that runs between Opportunity and Tekoa. This edition of the booklet was photographed and stripped by Robin Clayton. Additional typesetting was done by Teresa Ruggles who also designed the title page and colophon. Printing plates were made by Garry Adams, who also folded the sheets on a 26x40 Baum Dial-O-Matic folding machine. The sheets were printed by Trevor Del Medico using a Heidelberg press, Model KORS. Booklet binding was done by Garry Adams. The paper stock is 70 pound white offset. This was a fun project, we had no special difficulty with the work.